UNIVERSITY OF ILLINOIS
Agricultural Experiment Station

BULLETIN No. 249

CALCIUM CYANIDE FOR CHINCH-BUG CONTROL

IN COOPERATION WITH THE NATURAL HISTORY SURVEY,
ILLINOIS DEPARTMENT OF REGISTRATION AND EDUCATION

BY W. P. FLINT AND W. V. BALDUF

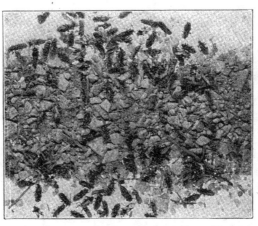

PORTION OF A CALCIUM CYANIDE STRIP USED AT RIGHT ANGLES TO A CREOSOTE BARRIER, SHOWING DEAD CHINCH-BUGS WHICH HAVE ATTEMPTED TO CROSS

URBANA, ILLINOIS, MAY, 1924

SUMMARY

Calcium cyanide is a substance having the appearance of slate, which on exposure to moisture gives off hydrocyanic acid gas. In sufficient quantities, it is deadly to all forms of plant and animal life.

The use of this chemical as a means of combating chinch-bugs was first tried in Illinois. Extensive tests were begun in 1922 and were continued thruout the season of 1923. In 1923 similar tests were made in a number of other states, particularly in Kansas, Missouri, and Indiana. While it is impossible to make definite recommendations as to its use from the results of two years' work in Illinois, it is believed that farmers will be warranted in trying out some of the methods described herein.

The best results from calcium cyanide have been obtained by using either the dust or the granules in combination with creosote or coal-tar barriers. Six-inch strips of the cyanide, requiring about one ounce to a strip, laid at right angles to the barrier every two rods, have under favorable conditions killed 75 to 95 percent of the bugs as they moved along the line of the barrier. This chemical was also very effective when dusted along the line of a coal-tar or creosote barrier.

Fair to good results have been obtained by dusting calcium cyanide in trap strips of crops sown between fields of small grain and corn. However, only one year's data have been obtained with trap crops.

Barriers of calcium cyanide alone were not so effective as the other methods and at the same time were more costly.

Hydrocyanic acid gas given off by calcium cyanide is very poisonous, and the directions for handling given on page 84 should be carefully noted.

CALCIUM CYANIDE FOR CHINCH-BUG CONTROL[1]

By W. P. FLINT, Entomologist, State Natural History Survey, and
W. V. BALDUF, Assistant Professor of Entomology,
University of Illinois

Calcium cyanide was first used in 1922 as a means of combating chinch-bugs around the margins of infested fields of small grain. Experiments were continued thru the season of 1923, a large number of tests being made during the latter season. In the first experiments this material was used alone as a barrier. The results showed it to be very effective, under some conditions killing every bug that came in contact with it, but its use in this way was so costly as to make it unpractical for general use on Illinois farms. Later experiments, however, showed that it could be used in combination with creosote and coal-tar barriers, thus greatly increasing their effectiveness without increasing the cost to a point where its use would be unpractical.

While definite recommendations of the best way to use calcium cyanide cannot be made from the results of two years' work, the results here presented will enable the Illinois farmer to judge for himself the best method of using this material on his farm.

In the Illinois experiments, four methods of using this chemical were tested:[2]

1. Laid in strips at right angles to a creosote or coal-tar barrier
2. Dusted or scattered along a creosote or other barrier at the time of day when the bugs are most numerous
3. Dusted over strips of trap crops sown between infested small grain and corn
4. Alone as a barrier around the margins of the infested field

---

[1] Note.—Calcium cyanide is a cheap form of rather low-grade cyanide. It has about half the strength of the sodium cyanide generally used in hydrocyanic acid gas fumigation of fruit trees, greenhouses, and flour mills. When this chemical is exposed to air containing the usual amount of moisture, or is applied to damp soil, the poisonous hydrocyanic acid gas is liberated in sufficient quantities to kill insects or other animals exposed to it, for varying periods of time.

[2] Some tests were also made to determine the effect of calcium cyanide dust when applied directly to corn. While it proved very effective in killing the chinch-bugs clustered on the plants, it cannot be used for chinch-bug control because it nearly always penetrates the curl of the leaves and causes the death of the plant or a severe burning. A 2-percent nicotine dust used in this way is just as effective as calcium cyanide dust in killing the bugs, is cheaper, and does not injure the corn.

Three forms of calcium cyanide were tried out: (1) flakes, which have the appearance of finely broken slate, the separate flakes being from one-fourth to one-half inch across and about one sixty-fourth inch thick; (2) granules, having the same form as flakes but being much finer, the larger particles being only about one-eighth inch across; (3) dust, the material ground to a fine powder.

USED IN STRIPS AT RIGHT ANGLES TO A CREOSOTE OR COAL-TAR BARRIER

In 1923 the method of control proving the most practical was the use of calcium cyanide strips laid at right angles to the creosote or coal-tar barrier. These strips are made by putting down one to one and one-half ounces of cyanide in each strip, laid from three to five inches wide and from six to eight inches long. These strips are made at right angles to the barrier, touching the creosote on the side next the stubble field. They should be spaced every two rods along the line. In the corners of the field, where the bugs usually gather in greatest number, it may be necessary to lay the strips closer together.

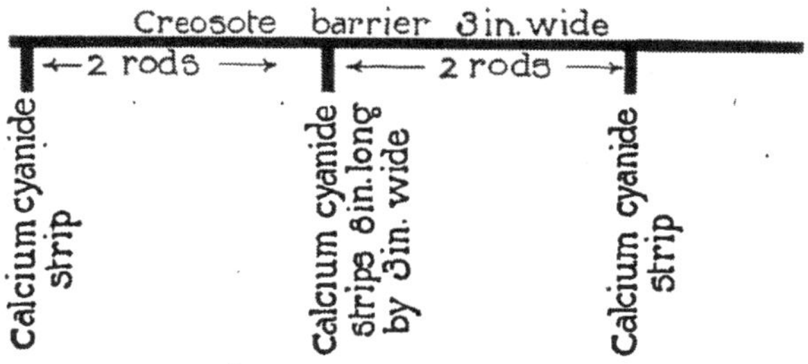

FIG. 1.—DIAGRAM SHOWING THE METHOD OF LAYING CALCIUM CYANIDE STRIPS AT RIGHT ANGLES TO A CREOSOTE OR COAL-TAR BARRIER

Care should be taken that the strips actually connect with the barrier. The slope toward the barrier should be gradual so that the cyanide will not roll down and away from the creosote or coal tar.

Chinch-bugs attempting to leave the fields of small grain and encountering barriers of creosote or coal tar are repelled by the odor of these substances. In the case of coal tar, they are held back by the stickiness of the material as well. Hence they turn and start along the stubble side of the barrier, seeking an opening thru which they can escape into fields of corn or to other food plants. Half-grown young, or nymphs, travel along the barriers at the rate of five to six feet a minute, and are almost sure to encounter a cyanide strip in five minutes or less after they get to the barrier.

Tests of this method extended over a period of two months in 1922 and a somewhat longer period in 1923, and were made under varying weather and soil conditions. Under some conditions pure dust was found to be the most effective form to use. When the soil is wet, the granules are better than the dust because they do not lose their strength so rapidly. The flakes also give better results than the dust when the ground is wet, but they are not very effective when the soil is dry or when the wind blows briskly. In general, granular calcium cyanide is the most practical form.

Care should be taken that the strips actually connect with the creosote barrier. The slope toward the barrier should be sufficiently gradual so that the cyanide will not roll down and away from the creosote.

A sprinkling can from which the sprinkling cap has been removed is a convenient vessel to use for pouring the cyanide. After a little practice one is able to quickly put down strips of approximately one ounce each at the intervals specified.

Strips Are Effective for a Limited Time

When the soil is dry and the wind strong, most bugs can cross the cyanide strips without being killed. Under more favorable conditions, however, the strips are effective up to five or more hours. Usually on warm bright days, the heavy movement of chinch-bugs starts between 2:30 and 3:30 in the afternoon, the bugs gathering along the barrier in greatest numbers between 3:30 and 5:30. The most effective kill can therefore be made by laying the strips between 3:00 and 4:00 o'clock in the afternoon. Except when the soil is dry and the wind high, strips laid at this time remain effective for the rest of the afternoon and may be counted on to kill 75 to 95 percent of all bugs reaching the barrier.

If the amount of cyanide in the strips is doubled, and the flake or granular form is used, some bugs will be killed the following forenoon, if no rain has fallen during the night; but the strips will not remain effective during the entire day. Since the bugs seldom migrate in large numbers in the morning, the kill is not sufficiently increased to warrant the added expense of using these larger amounts of cyanide.

Occasionally on cloudy days there is a considerable movement of chinch-bugs all day. In such weather the cyanide should be applied between 9:00 and 10:00 o'clock in the forenoon, and another application made in the afternoon if necessary. *Because the creosote or coal-tar barrier holds the chinch-bugs in the field, it is very seldom necessary to make more than one application a day, even during cloudy weather.*

EXPENSE OF COMBINED BARRIER

The total cost for a quarter of a mile of creosote or coal-tar barrier and cyanide strips along it is about $21 for the migration period. This is figuring cyanide at 20 cents a pound delivered at the local shipping point and the cost of creosote or coal tar at $5 to $7 for the season. This will not be thought too expensive to be practical if one remembers that he is not only stopping the bugs but is actually killing the majority of them. Post holes are unnecessary where calcium cyanide cross-strips are used, for the cyanide kills more bugs than are usually caught in the post holes. The saving of the labor of digging the post holes helps pay the expense of the cyanide.

USED ALONG A CREOSOTE OR COAL-TAR BARRIER

The second method of using calcium cyanide in combination with a creosote or coal-tar barrier is to distribute small amounts along the barrier, on the side of the field of small grain, instead of in strips at right angles to the barrier. To be effective, the calcium cyanide should be laid down each day when the bugs are gathering in greatest numbers. For this method, the cyanide dust has been found most practical.

The expense of this treatment is somewhat higher than that of the right-angle strips, but on windy days, when the soil is very dry, it is more effective and more convenient than the strips. The cost of the calcium cyanide is from $1 to $1.50 a day for each quarter-mile of barrier.

This method has the disadvantage of requiring a close watching of the barrier, for in order to get a maximum kill the dust must be applied when the bugs are present along the barrier in greatest numbers. While this is usually about 4:30 in the afternoon, the time may vary a great deal with weather conditions. One application a day is necessary. This kills all the bugs massed against the barrier at the time and those coming into the dusted area for ten to thirty minutes thereafter. The duration of the kill depends upon weather conditions.

HOW TO APPLY THE DUST

The calcium cyanide dust may be applied with a blower duster or a homemade shaker. The duster is of the type commonly used for dusting truck crops or small fields of cotton; it costs from $10 to $25, and is apparently no more effective for this purpose than a homemade shaker. A convenient and efficient shaker can be made in the following way:

Obtain a gallon or half-gallon tin pail with a tight-fitting lid. (The lid is important in confining the fluffy and poisonous dust.) Make an extension bail by nailing narrow strips of wood eighteen inches long to opposite sides of the pail and joining them above with a piece of broom stick. (The extension bail enables the operator to walk erect and at the same time to distribute the material close to the ground so that it will not be carried away by the wind.) From the outside of the pail, punch four to six holes per square inch in the bottom, using an eight-penny nail.

FIG. 2.—HOMEMADE SHAKER FOR APPLYING CALCIUM CYANIDE DUST
This shaker must have a tight-fitting lid to prevent the escape of the poisonous dust.

Fill the pail two-thirds full of dust and walk along the barrier, forcing the material out by a jiggling motion. The dust may be applied as rapidly as one can walk down the side of the field. Approximately one pound of the dust to thirteen rods of barrier was found to be effective.

DIRECTIONS FOR MAKING A CREOSOTE BARRIER

A creosote barrier is made by throwing up a ridge of earth around the infested field of small grain and applying creosote along the brow of the ridge on the side toward the field of small grain. The furrow should be turned away from the grain field.

The creosote is applied with a galvanized or tin bucket having a six-penny nail punch in the side near the bottom. The stream of creosote should be directed against the brow of the ridge on the side toward the field of small grain, as above stated, so that the bugs in attempting to leave the field will crawl up to it. The creosote must be renewed once a day during the first week, and should always be applied on the same line. After the first week an application every other day is usually sufficient.

The application should be made usually about 1:00 o'clock in the afternoon, before the bugs have gathered along the line. It is the odor

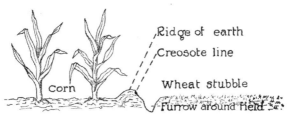

Fig. 3—Construction of a Creosote Barrier Used Alone
Where calcium cyanide strips are to be used with the creosote or coal-tar barrier, the slope of the ridge should be more gradual than shown here.

of the creosote that keeps them from crossing the barrier; consequently if the line is renewed when the bugs are massed against it, many become confused by the strong smell of creosote in the air and run over the barrier.

A barrel of creosote, on an average, is required to maintain a half-mile of barrier during the period the bugs are leaving the wheat stubble. The creosote should preferably be of a grade with a high napthalene content.

A line of ground limestone is sometimes spread along the brow of the ridge and the creosote applied on this line. Limestone retains the odor of creosote better than does the soil and so increases somewhat the effectiveness of the barrier.

DIRECTIONS FOR MAKING A COAL-TAR BARRIER

Coal tar may be used as a barrier in much the same manner as creosote, but the ground on which it is poured should be nearly level and should be compacted as firmly as possible, and a larger quantity of coal tar than of creosote will need to be used.

It must be borne in mind that the effectiveness of the coal-tar barrier depends largely on its sticky character. It will have to be renewed more frequently than the creosote, and since about twice as much coal tar as creosote is required, this kind of barrier usually costs more than one of creosote.

Coal tar for this work may be obtained from gas plants in any of the larger cities. There is a wide difference in the grades that may be obtained from different plants. Coal tar which contains practically all the creosote is much more effective for barriers than that from which a part of the creosote compounds have been distilled.

USED ON TRAP CROPS

Many attempts have been made to control chinch-bug migrations by stopping the bugs in narrow trap strips of their favorite food plants sown between the small grain fields and the fields of corn or other crops to be protected. There seemed to be a possibility of using calcium cyanide with such trap crops by applying it when the bugs were gathered in the trap strips, and in 1922 some experiments were made with this method. No special trap crops had been sown, but strips of wild grasses growing between the small grain and the corn were dusted with calcium cyanide and a fair kill of bugs was obtained.

In 1923 this trap crop and cyanide combination was tested more thoroly. Strips of oats, millet, Sudan grass, and corn were sown between heavily infested fields of small grain and of corn. These strips were treated with flake, granular, and dust calcium cyanide at varying intervals and at varying rates of application. Altho these tests extended over the migration period of one season, the results were not conclusive. Apparently this method is not so effective as the creosote barrier with the cyanide laid along it or in strips at right angles to it. However, the method may be worth trying under some conditions.

Of the trap crops used on the University Farm in 1923, Sudan grass was decidedly the most effective. It is very attractive to chinch-bugs and withstands a considerable amount of feeding by them before it is killed. It is vigorous, and hence can also endure the bruising by teams and machinery which occurs in cultivating the corn. It is not easily killed by heavy applications of calcium cyanide, and the sheaths of the leaves fit so tightly to the stems that chinch-bugs cannot get between them and the stalk and thus be protected from the fumes of the gas. The rows of Sudan grass should be sown thickly and close together so that a bug can scarcely pass thru the strip without encountering a plant.

Some tests with rye made during the latter part of the summer showed that this plant is apparently second to Sudan grass as a trap crop. It is hardier than oats and nearly as attractive to the bugs as Sudan.

Millet and oats were found to be quite susceptible to cyanide.

Corn, even when seeded with a drill, formed a very poor trap crop, many bugs going thru the strip and into the field of corn beyond. The corn was also very susceptible to the effects of the cyanide dust, one or two dustings killing the plants.

From the results of this work thus far, the following statements can be made:

To be in the best condition for stopping the bugs, a trap crop should form a thick, close covering over the ground and should not be more than six inches high at the time the chinch-bugs migrate. For this reason, the trap strip should not be sown when the corn is planted,

FIG. 4.—DUSTING A TRAP-CROP STRIP WITH A BLOWER DUSTER

The trap crop shown here is Sudan grass, which is somewhat higher than it should be for this purpose. To get the best results, the trap crop should not be more than six inches high.

unless rye is used. Sudan grass should be planted one to two weeks after the corn is in the ground. The time of planting depends, however, on the expected time of small grain harvest and not on the size of the corn. The trap strips should be three and one-half feet or more wide, and to insure a thick growth, they should be sown with about one and one-half times the usual amount of seed.

In this method of using calcium cyanide, dust was found to be the most desirable form.

The dust may be applied to the trap crops by means of a duster or a homemade shaker. The shaker is not so good for this purpose as the regular blower duster; however, it gives a very fair distribution of the dust providing the trap strip is not more than six inches high. The trap strips must be dusted once a day when a heavy migration of the bugs is taking place. The dusting is most effective when done between 4:30 and 5:30 in the afternoon.

Judging from experiments made thus far at this Station, calcium cyanide dust must be applied at the rate of about one pound for each two and one-half rods of trap strip three feet wide. This is at the rate of one pound of dust to each one hundred and twenty square feet. Under favorable conditions, such an application kills from 80 to 95 percent of the bugs gathered in the strip, but when the wind is strong and the crop dry, the results are not nearly so good. The amount of kill seems to depend in a general way on the amount of moisture present in the trap crop, the velocity of the wind, the height and density of the trap strip, and perhaps some other factors.

Calcium cyanide granules may be used in treating these trap strips and can be applied by hand at the same rate as the dust. Where the trap crop is wet with rain or dew, the granules are seemingly fully as effective as the dust, but on dry soil and plants they do not result in so large a kill.

The above statements are not to be taken as definite recommendations for the use of the trap strip and calcium cyanide combination for chinch-bug control. It will be necessary to carry on experiments for one more season at least before definite recommendations can be made.

USED ALONE AS A BARRIER

More than three hundred separate tests have been made in Illinois to determine the practicability of the flake, granular, and dust forms of cyanide used alone as chinch-bug barriers. Under favorable conditions it was found that nearly all the bugs attempting to leave a field of small grain on foot may be killed along such a barrier, if a sufficient amount of the cyanide is applied. Under less favorable conditions, however, the results were extremely variable. In some kinds of weather, one-half to one-fourth pound of cyanide per rod of barrier killed all bugs attempting to cross it during a period of three to five hours. With other weather conditions, the same amount permitted 50 percent of the bugs to cross in fifteen to thirty minutes after it was applied.

In these tests, the calcium cyanide was used in various positions and at various rates. It was used on soil from which all vegetation had been removed, on the top of the plow furrow ridge, in strips in the stubble (which protected it somewhat from the action of the wind and sun), and in the bottom of a plowed furrow or trench dug around the margins of the field. Only when used in a furrow or trench did it give promise of having practical value as a barrier, under the conditions encountered. Used in this way the gas is somewhat confined, and if the soil is damp and little or no wind blows down the furrow, the gas accumulates sufficiently to kill most of the bugs over a considerable period of time. If the soil is dry, and the wind is strong along the furrow, the kill is small, even when the cyanide is used in the bottom of the furrow.

In Table 1 are shown the results of 144 tests of cyanide used in the furrow; they give an idea of the diverse effects of the chemical under different weather conditions encountered.

TABLE 1.—EFFECTIVENESS OF VARIOUS AMOUNTS AND FORMS OF CALCIUM CYANIDE USED ALONE IN THE FURROW AS CHINCH-BUG BARRIERS

Form	Amount used on each rod of barrier	Number of tests	Average time during which 90 percent of bugs were killed	Variation in time during which 90 percent of bugs were killed
	lbs.		hrs.	hrs.
Flakes.....	4	3	8¾	3 to 20
Flakes.....	2	3	8½	1 to 17
Flakes.....	1	18	4	½ to 15
Flakes.....	½	15	1½	½ to 5½
Flakes.....	¼	17	1¼	¼ to 4
Flakes.....	⅛	4	¾	¼ to 1½
Granular...	2	3	4¾	1 to 10
Granular...	1	15	3¼	½ to 17
Granular...	½	12	1¾	½ to 5
Granular...	¼	13	1½	½ to 6
Dust, pure .	4	1	5	......
Dust, pure .	2	1	5	......
Dust, pure .	1	13	3½	1 to 5
Dust, pure .	½	11	2½	½ to 6
Dust, pure .	¼	10	1¼	½ to 3
Dust, pure .	⅛	5	¾	½ to 1

On still, damp days, one-half to one-fourth pound of flakes, granules, or dust of calcium cyanide to a rod makes an ideal barrier in a furrow five to seven inches deep. The cyanide should be applied on the furrow bottom at the point farthest from the grain stubble. The bugs are then forced to climb up the steep side of the furrow directly above the cyanide, and as a result, their movements are slowed up. Thus they are exposed to the gas for a longer time than when the cyanide is placed in the middle of the furrow or along the side down which they tumble in leaving the stubble.

As may be seen from Table 1, where one-fourth to one-half pound of calcium cyanide per rod was used, it was in some cases effective for as long as six hours. Under such conditions, the cyanide applied at 1:00 o'clock could be depended upon to kill 90 percent of all the bugs

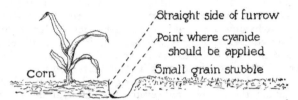

FIG. 5.—MOST EFFECTIVE CONSTRUCTION OF A CALCIUM CYANIDE BARRIER
The straight side of the furrow should always be toward the corn, and the calcium cyanide should be laid at the base on this side, as shown above. If a trench is dug, the cyanide should be applied at the same point.

attempting to leave the field during the afternoon. In other cases, the same amounts of cyanide applied in the same way and in the same places were effective only for fifteen to thirty minutes; that is, it would be necessary to make three to five applications to maintain an effective barrier during an afternoon.

Used at the rate of one-fourth of a pound per rod, an amount of calcium cyanide sufficient to maintain a barrier a quarter of a mile long would cost $4 a day. This figure is based on a price of 20 cents per pound for the cyanide delivered at the farmer's shipping point. For a fourteen-day migration period, material for a quarter-mile of barrier would cost $56. The experiments show that on a number of days two or three times this amount of cyanide would have to be used; therefore one can hardly count on maintaining an effective barrier for the season at less than $70 to $75 for each quarter-mile. This is nearly three times the expense of killing the bugs with a combination of creosote and calcium cyanide, or nearly ten times the cost of maintaining a creosote barrier.

Because of the great variation in the results of experiments made to date, definite recommendations for the use of cyanide alone as a barrier cannot be made at this time. During the past season work similar to that done in Illinois has been carried on in Kansas, Missouri, and Indiana. Rather favorable results have been obtained with the barriers in Missouri. The results in Kansas and Indiana compare very closely with those obtained in Illinois. Judging from all the tests made, it may be said that a chinch-bug barrier can be maintained by the use of calcium cyanide alone, but that the expense of maintaining such a barrier is so great as to make it unpractical except for the protection of a very valuable crop, such as seed corn.

Of the different forms of calcium cyanide tried for barrier work, the dust has, on the whole, given the best results, mainly because of its physical character. The small particles afford unstable footing for the bugs when they attempt to cross the line. This retards their movement, and they are exposed to the gas for a considerably longer time than when trying to cross a line of flakes or granules. These coarser forms of the cyanide evolve the gas more slowly than the dust, and do not greatly retard the movements of the bugs. On the other hand, when the soil is very damp, the dust gives off its gas so rapidly that it soon weakens and does not continue to kill over a very long period. Under such conditions the coarser forms of cyanide are to be preferred.

HOW TO HANDLE CALCIUM CYANIDE

Caution is necessary in handling calcium cyanide. The gas given off from this material is one of the most deadly known. However, the calcium cyanide used for chinch-bug control is comparatively weak in hydrocyanic acid gas, and if one is careful not to get his head directly over the container, or to breathe in the dust or the fumes of the freshly opened cans, there is little danger. There is apparently very little hazard in handling the material in the open air if one keeps on the windward side of the container. The writers have worked with it for the past two seasons, and have purposely tested the effects upon themselves in a number of different ways. Occasionally when enough of the dust was breathed into the nostrils, particularly when the distributing utensils were being filled from the larger cans in which the material was shipped, a severe headache was caused. This discomfort disappeared after an hour or two in the open air; but one must be careful at all times. *Cyanide should not be left in the field or around buildings where children or farm animals might possibly reach it.* Fowls that have access to it sometimes pick up the flakes or granules, with fatal results.

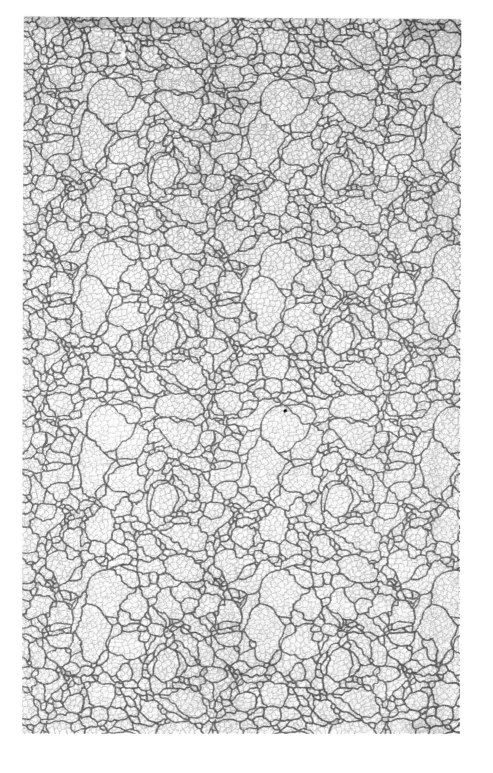

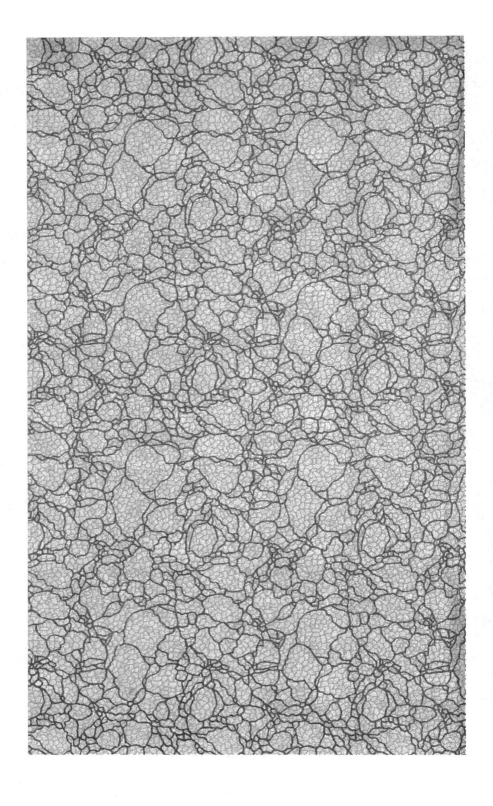

UNIVERSITY OF ILLINOIS-URBANA
Q.630.7IL6B   C002
BULLETIN. URBANA
246-256 1923-25

3 0112 019529103

Printed in the USA
CPSIA information can be obtained
at www.ICGtesting.com
LVHW050927120524
780060LV00008B/143